Home on the Range

— THE STORY OF THE — NATIONAL BISON RANGE

Published by

Helena & Billings, Montana

Published by Falcon Press Publishing Co., Inc. in cooperation with the Glacier Natural History Association and the U.S. Fish and Wildlife Service.

Library of Congress Catalog Card Number: 84-73222

ISBN: 0-934318-51-4

Distribution and Marketing: Falcon Press, P.O. Box 279,
Billings, MT 59103

Editorial and Production: Falcon Press, P.O. Box 731,
Helena, MT 59624

Design, typesetting and all other pre-press work
by Falcon Press in Helena, Montana.

Printed in Hong Kong.

Editor: Sally K. Hilander
Design: DD Dowden
Text: Jon Farrar

This book is dedicated to my friends at the National Bison Range, both to the U.S. Fish and Wildlife Service employees, including those who have moved on, and to the animals I became acquainted with because of their cooperation.

Danny On also helped make this book possible. Danny, pictured left, encouraged me to try my hand at wildlife photography. Our first trip was to the National Bison Range in 1970. This book is partly his. Thanks Danny, Jon Cates. Photo by Jon Cates

Jon Cates lives near the National Bison Range, probably the main reason he became interested in wildlife photography and switched over from stalking wildlife with rifle to a camera. Cates, pictured above, has had photos on the cover of most of the national outdoor magazines and in numerous published articles. Photo by Danny On

A buffalo family roams the sweeping expanses of the National Bison Range, against the spectacular backdrop of the Mission Mountains. About 100,000 of the giant beasts are alive today, remnants of a species that once numbered thirty million and roamed most of the North American continent. Michael Sample

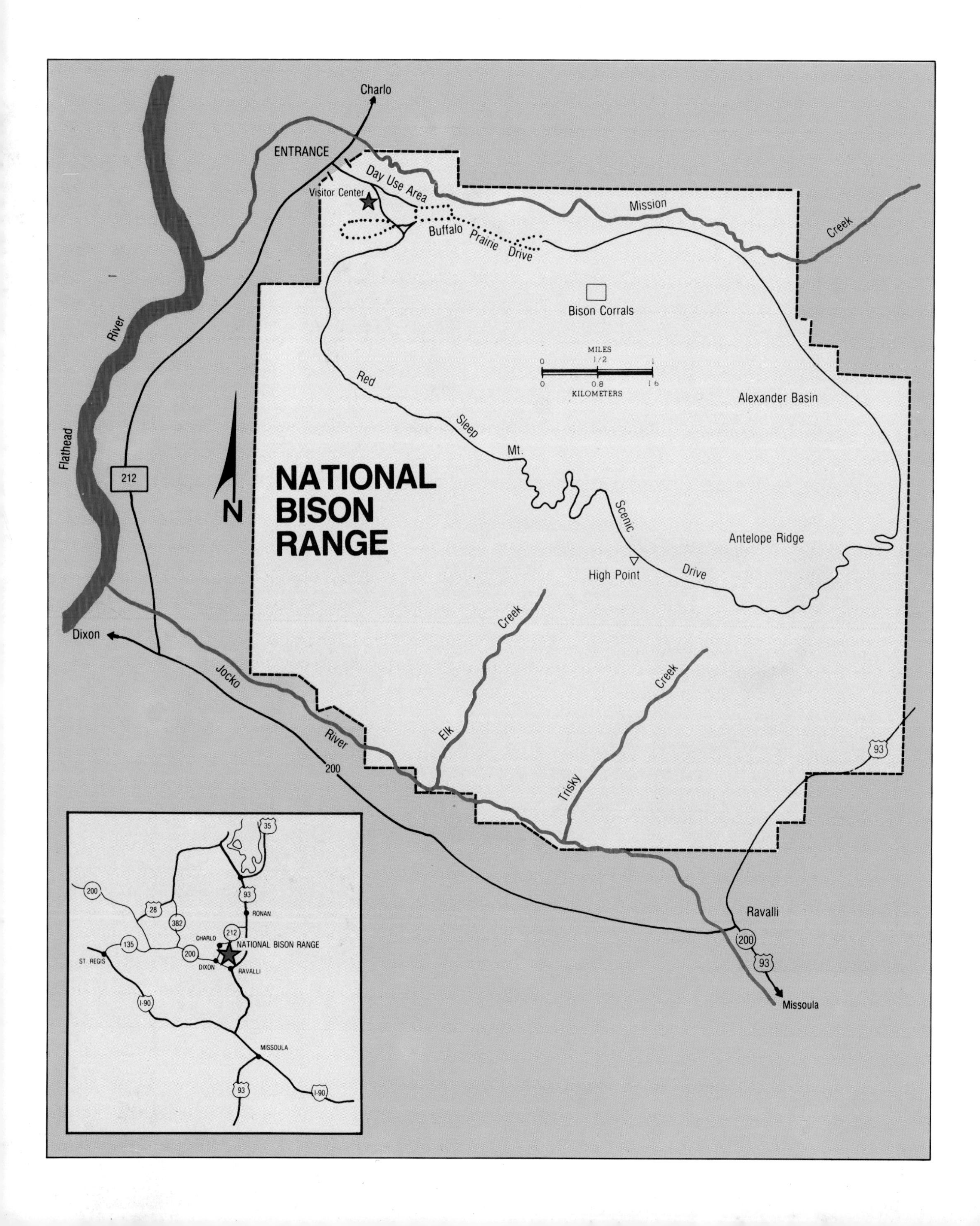
Charlo
ENTRANCE
Day Use Area
Visitor Center
Buffalo Prairie Drive
Mission Creek
Bison Corrals
MILES
0
1/2
1
0
0.8
1.6
KILOMETERS
Alexander Basin
Red Sleep Mt. Scenic Drive
Flathead River
212
N
NATIONAL BISON RANGE
Antelope Ridge
High Point
Elk Creek
Trisky Creek
Dixon
Jocko River
200
93
Ravalli
200
93
Missoula
35
200
93
RONAN
28
382
CHARLO
212
135
NATIONAL BISON RANGE
ST REGIS
200
DIXON
RAVALLI
I-90
MISSOULA
93
I-90

INTRODUCTION

Visitors take a journey into the past when they enter the National Bison Range. Jon Cates

The pristine 'palouse prairie' is a blend of grassland and mountains, unique from the Great Plains east of the Continental Divide. Michael Sample

When Europeans penetrated the interior of the North American continent, they encountered herds of buffalo that were days in passing and extended as far as the eye could see.

During the 1870s and 1880s, hide hunters, with the de facto support of a government intent on subduing the Plains Indians, methodically slaughtered the seemingly limitless herds. By 1889 it was estimated that fewer than one thousand buffalo remained, and only eighty-five of those were free-ranging. This symbol of the American West teetered on the edge of oblivion.

From this black episode in history emerged the National Bison Range, the child of a nation's conscience. It was created by an act of Congress in 1908 as the second refuge specifically for preservation of the buffalo.

The National Bison Range is a symbol of our national heritage, a slice of America's untamed past, where life goes on for the native wildlife much as it has for thousands of years. A rush of wildflowers still races across the valleys in spring. Newborn pronghorns cavort on wobbly legs in the basins, and the western meadowlark's gurgling song announces the morning. Age-old dramas are played out much as they have been since the last glacier retreated some ten thousand years ago: Bighorn rams clash horns for possession of ewes, short-eared owls harvest the overabundance of deer mice born each summer, and drab little grassland sparrows build their nests on the prairie soil.

Both the harsh reality and the beauty of nature are evident on the prairies and in the woodlands of the National Bison Range. Here the intricate workings that drive a natural system mesh in perfect synchronization.

Nestled beneath the snow-capped peaks of the Mission Range, near the tiny town of Moiese on the Flathead Indian Reservation in western Montana, the National Bison Range is a unique blend of grassland and mountains.

For the patient listener, the National Bison Range has many stories to tell. In its rocky outcrops are recorded the turbulent eras when great blocks of the earth's crust were thrust up, when time was marked by the ponderous advance and retreat of glaciers. Tales of mountain men and warring Indian tribes, mountain lions and grizzly bears, whisper through the ponderosa pines. In the grasslands are written lessons of ecology, of the interdependence of life.

The National Bison Range is more than a buffalo refuge, more than a museum where the wilderness has been preserved. It is a touchstone by which we can learn who we have been, what we have become, and what we can be.

Rocky Mountain bighorn sheep prefer the higher reaches of the National Bison Range. Jon Cates

The American bison, dubbed 'buffalo' by European settlers, is the largest land mammal in North America, with mature bulls weighing as much as 2,500 pounds. In confrontations, they arch their humped backs, thus enhancing their already formidable presence. Jon Cates

Alert to possible danger, skittish white-tailed does raise their 'flags' and prepare to bound off across the prairie. Whitetails are renowned for their ability to outsmart hunters. Jon Cates

The buffalo escaped extinction when early-day conservationists recognized them as a symbol of our national heritage and fought to save them. Michael Sample

Flexibly adapted to nature's whims, buffalo drink enormous quantities of water when they find it, but they also can go for several days with none. Plainsmen believed the enormous beasts could smell water five miles away. Michael Sample

Perhaps no other animal is as competent and comfortable in lofty heights as the Rocky Mountain goat with its suction-cup hooves. The name is a misnomer. The animals are not true goats, but rather part of a goat-antelope family that includes the Alpine chamois. Billies may stand three and a half feet at the shoulder and weigh 250 pounds. Jon Cates

Bull elk sometimes stand five feet at the shoulder, weigh from 600 to 800 pounds, and sport antlers that span four feet. Antlers are shed each winter, after the battles of the rutting season. New growth begins in March and the antlers are polished to a hard shine by August. Jon Cates

One of the 187 bird species that call the National Bison Range home, the ring-necked pheasant was introduced from Asia and is now common from Canada to the Texas panhandle. Jon Cates

The columbine is an early bloomer, beating the competition for moisture that becomes scarce as summer visits the prairie. At least nine species occur in North America, but all are becoming rare, in part because of overgrazing. Jon Cates

Palouse prairie grasses are wonderfully adapted to their environment, with long, slender, supple leaves that conserve moisture and employ the wind to scatter their seeds.
Michael Sample

A NATION'S SHAME

'Buffalo runners' scoured the plains after the Civil War in a senseless extermination that left as many as five carcasses rotting for every hide shipped, much to the dismay of the Plains Indians to whom the great animals were a living commissary. A strong herding instinct made the bison vulnerable to mass slaughter because they refused to scatter.
Montana Historical Society

America's vast buffalo herds seemed inexhaustible when encountered by European explorers on the Great Plains. Francisco Vasquez de Coronado described them in 1540:

". . . I reached some plains so vast that I did not find their limit anywhere that I went, although I traveled over them for more than 300 leagues (about 900 miles). And I found such a quantity of cows. . . that it is impossible to number them, for while I was journeying through these plains, until I returned to where I first found them, there was not a day that I lost sight of them."

Buffalo once roamed two-thirds of the North American continent, from most of the Atlantic coast westward across the Rocky Mountains; from the Gulf of Mexico northward into the Canadian provinces of Alberta, Saskatchewan, and Manitoba.

The number of buffalo that originally inhabited the continent has been a subject of debate since the great herds were first encountered. Early accounts employed unquantitative description such as "teeming myriads." Estimates from a century ago placed the number in America at sixty to seventy million. A more recent estimate, based on the number of buffalo that could have co-existed with other native grazing animals on the available range, is thirty million.

The killing began in the early 1700s, wherever settlers encountered buffalo. Indian tribes along the upper Missouri River tanned winter robes to barter with at trading posts. White settlers killed the giant beasts for meat and to protect their crops. But this harvest, which continued until about 1830, had relatively little impact.

During the 1860s, railroads surged across the mid-continent, but not until 1871 did commercial tanners learn how to process the buffalo's thick hide. That discovery set the stage for the most devastating slaughter of a wild species the world had ever witnessed.

During the fall of 1871 and winter of 1872, hide hunters poured across the plains of southern Nebraska, western Kansas, and eastern Colorado. Each hunter took as many buffalo in a day as his crew of skinners could handle, occasionally dropping more than a hundred from a single stand. Prodded by greed for hides that brought one to three dollars apiece, thousands of "buffalo runners" scoured the plains. For every hide shipped, perhaps five carcasses lay rotting.

Buffalo chaps, a turn-of-the-century 'novelty fur,' are modeled by Charles Allard, Jr. The tough hide, of course, also served a practical purpose. Montana Historical Society

The southern plains were quickly stripped of buffalo. Treaties and the powerful Sioux nation only temporarily blocked exploitation of the "northern herd" that roamed Nebraska, Wyoming, Montana and the Dakotas.

In 1881, the Northern Pacific Railroad arrived at Miles City, Montana, and provided a shipping point for hide hunters. The slaughter was both efficient and short-lived. Hunting outfits pushing across the Montana plains in the fall of 1883 encountered not a single buffalo herd.

Just as scavengers followed prairie wolves, bone pickers trailed the hide hunters. Bleached skulls and skeletons, grim reminders of a nation's callousness, were piled into enormous ricks at railsides and shipped east to be carbonized for use in purging raw sugar of its brownish color or ground into fertilizer. Even horns and hooves were gathered to become buttons, combs, knife handles, and glue.

East of the Mississippi, the buffalo were essentially gone by 1800. In 1832, plains artist George Catlin predicted that "the buffaloe's doom is sealed." Naturalist and painter John J. Audubon wrote in 1843: "Like the Great Auk, before many years the Buffalo will vanish. Surely this should not be permitted?"

Legislation to save the buffalo was thwarted by a government that viewed the slaughter as a means of subjugating the Plains Indians, who were dependent on the animals for food, clothing, tools, weapons and shelter.

In 1889, William Hornaday, whose name is closely allied with the effort to save the buffalo, estimated that there were no more than 85 free-ranging in the United States, 200 protected by law in Yellowstone National Park, 550 near Canada's Great Slave Lake, and about 250 in zoos and private herds.

Statistics from the Northern Pacific Railroad tell the story: In 1882, 200,000 robes were shipped; in 1883, 40,000; in 1884, one carload; in 1885, none. Montana Historical Society

'Big Medicine,' an albino buffalo bull that was a legend to the Indians, roamed the National Bison Range for twenty-six years. Montana Historical Society

Members of the northern herd grazed peacefully on the Montana prairie before hide hunters arrived by the thousands to exterminate them. Montana Historical Society

Wallowing provides buffalo with relief from flies and other insects that become imbedded in their dense hair. By lying down and kicking dust over themselves with their forelegs, they are able to smother the pests to death. Jon Cates

Montana's state flower, the bitterroot, is named Lewisia rediviva in honor of Capt. Meriwether Lewis, who noted it in 1906. Indians still collect the nutritious, fleshy root, which is baked, boiled or dried and ground into meal. Michael Sample

Pronghorns have stiff, air-filled body hair that buffers them from the cold. Rump hair is extra long and a network of muscles erects it when the animal is alarmed. This creates a large white shield that reflects light and alerts other pronghorns to possible danger. Jon Cates

The bandit-faced raccoon, which inhabits all 48 contiguous states and Canada, finds the forested areas of the National Bison Range an excellent choice for easy living. Alan Nelson

An American kestrel or 'sparrow hawk' huddles in his feathers as the temperature dips to twenty-eight below zero. Smallest of the North American falcons, the kestrel prefers hunting in open country, scanning the terrain from a perch. Jon Cates

SAVING A NATIONAL HERITAGE

Michel Pablo ran a buffalo ranch in the Flathead Valley until 1906, when he sold his herd of 600 to Canada, which was anxious to stock its new national park at Banff. For $200,000, he agreed to deliver the herd by rail to Wainright, Alberta, twelve hundred miles away. Montana Historical Society

The extraordinary roundup took six years and captured the attention of writers, photographers and artists like Charles M. Russell, standing at center. Russell was disgusted that the U.S. government had left it up to the Canadians to save the endangered buffalo from extinction. Montana Historical Society

The beginning of the twentieth century was the turning point. Newspapers and magazines were filled with articles encouraging the protection and re-establishment of the buffalo. In 1905, the American Bison Society was formed, with Theodore Roosevelt named honorary president and William Hornaday elected president. As a result of the society's lobbying in Congress, the 59,000-acre Wichita Mountains Wildlife Refuge in southwestern Oklahoma was founded and stocked with fifteen animals from the New York Zoological Society.

Creation of the National Bison Range did not officially occur until 1908, but symbolically it began in 1872 with a Pend d'Oreille Indian named Samuel Walking Coyote. The story has variations, but a popular version is that a few orphaned buffalo calves trailed him across the Continental Divide when he returned to the Flathead Valley from a traditional hunting trip to the eastern plains. The herd increased to thirteen by 1884, and was sold to local ranchers Charles Allard and Michel Pablo.

When Allard died in 1896, the herd numbered three hundred and was the largest in the United States. His share of the buffalo went to heirs, and nine years later, Pablo offered to sell his herd to the federal government. President Roosevelt tried without success to convince Congress to establish a buffalo refuge in Montana, and Pablo instead sold his herd to the Canadian government.

With support of the American Bison Society, Montana Senator Joseph M. Dixon introduced a bill to the 1908 Congress to establish a buffalo range in his home state. Time was running out. Land on the Flathead Indian Reservation had been opened to homesteading. At the last minute, the measure was attached to an agricultural appropriations bill that sped through Congress.

Finally, on May 23, 1908, legislation creating the National Bison Range was signed. Land was purchased from the Salish, Kootenai and Pend d'Oreille tribes for about $40,000 and placed under management of the Bureau of Biological Survey, later to become the present-day U.S. Fish and Wildlife Service. The American Bison Society raised $10,000 to purchase a nucleus herd of thirty-seven buffalo, descendants of Samuel Walking Coyote's orphaned calves.

Michael Pablo's mounted cowboys, led by foreman Charles Allard, Jr., center. Montana Historical Society

Buffalo were hauled in special reinforced wagons to the trains as Pablo's roundup began to show signs of success. Some escaped, some were killed and more were born. About seven hundred were finally delivered live.
Montana Historical Society

The most cantankerous bulls, those with a habit of charging humans, were shot during the roundup, but Indians saw to it that hides and meat weren't wasted. This photo was taken in 1907.
Montana Historical Society

Among the buffalo rounded up for shipment to Canada was Old Stub Horns, a notorious bull believed to be thirty years old.
Montana Historical Society

On October 17, 1909, the National Bison Range became a reality. William Hornaday reported: "As the crates were opened, the animals backed out of them, looked about for a moment, saw their Paradise Regained looming up on the farther side of the Jocko River, splashed across the stream, and climbed into their new home."

The 18,542 fenced acres were indeed paradise: Douglas fir and ponderosa pines capping the upper hills; thick stands of alder, juniper, aspen, birch, cottonwood, and willows crowding the creek and river bottoms; and, most attractive of all to the buffalo, lush intermontane grassland.

In such abundance, the fledgling herd prospered, rapidly filling its role as the dominant species of the grassland community. Other grazing animals were introduced to fill vacant niches in the Bison Range's diverse environment: Pronghorn antelope and white-tailed deer in 1910; elk in 1911; mule deer in 1918, and bighorn sheep in 1922.

The paradise soon was tarnished. The buffalo's brush with extinction only a decade earlier weighed heavily on the nation's conscience, so the goal at first was to increase their numbers at any cost. Predators such as coyotes, bobcats and golden eagles were trapped, shot, and poisoned to ensure the survival of the buffalo calves and the young of other grazing animals.

With little competition for forage, the numbers of buffalo and elk spiraled upward. By the mid-1920s, the Bison Range held nearly 700 buffalo and 650 elk. With most of the predators eliminated, rodent populations mushroomed and competed for a dwindling food supply.

Large reductions of animal numbers followed and a supplemental feeding program was initiated to carry the remaining buffalo and elk through the winters, and ease grazing pressure. Such remedies forestalled a massive die-off, but for other species they came too late. In 1926, the entire herd of sixty pronghorns died, probably as a result of their weakened condition and vulnerability to disease.

By the early 1930s, the number of elk and buffalo had been reduced by half. Construction of watering ponds helped distribute grazing pressure over the range, but this was not enough. Internal fencing was needed and this job was begun, but the nation was in the grip of the Depression, and funding was not available to complete it. The National Bison Range remained three huge pastures in which the buffalo continued to concentrate and overgraze. In 1941, supplemental feeding was discontinued because it was both costly and unsuccessful at improving the range.

Unaware that this is his ticket to freedom, one of Pablo's bulls is desperate to escape. Montana Historical Society

One of the Pablo herd leaves its home turf for Canada and freedom. Montana Historical Society

The refuge entered the 1950s on the brink of ecological collapse. But the fledgling sciences of range and wildlife management were becoming more sophisticated, providing the knowledge to bring the plant and animal communities back into balance. The range was fenced into eight pastures to better regulate grazing, and buffalo herds were systematically cropped to maintain a healthy mix of bulls and cows of different ages. Coyotes and other predators were recognized for their role in suppressing rodent populations that compete with large grazing animals for forage.

By the early 1960s, the range was beginning to show signs of recovery. Undesirable weeds were crowded out by a healthy mix of native grasses and forbs. Pronghorns had been re-introduced in 1951. In 1964, mountain goats were added. The National Bison Range became a center for range and wildlife management research, providing answers to the complex questions of how to manage a natural environment.

During the mid-1960s, range management techniques were honed to a fine edge. A deferred grazing rotation system was introduced. The buffalo herd was divided into two smaller herds, one numbering about 180 and the other about 130. Each herd is rotated among four of the eight pastures at regular intervals. The pastures that were grazed during the previous growing season are skipped, thus ensuring that no pasture is grazed during the same period year after year, and only once every four years during the critical spring growing season.

Each October, the U.S. Fish and Wildlife Service steps in to cull the herds, thus keeping the number of animals in balance with the habitat. The entire population of buffalo is gathered by horseback from the range and worked through roundup corrals, where they are checked for illness and the young females vaccinated against brucellosis, a disease that can cause abortion. Calves are branded with their birth year.

After this process is complete, 300 to 350 buffalo, the number wildlife managers have determined the National Bison Range can support, are released. Bulls are alternated between the two herds to reduce inbreeding. The remaining seventy-five or eighty buffalo from all age classes are delivered live to buyers who have purchased them through sealed bids.

Selective harvest by range officials keeps mule deer numbers at 200 to 250, and whitetailed deer at 150 to 200. The meat is donated to area schools. Live trapping and relocation accomplishes the goal of maintaining 100 to 150 elk, 100 to 120 pronghorns, 50 bighorn sheep and 40 mountain goats. Predators are allowed to roam the range, although coyote numbers are kept at 25 or 30.

But why manage at all? The grassland community was working well before man interfered. Why not let nature work out the proper balance between plants and animals?

The National Bison Range is not a truly natural environment. The fence that contains the large grazing animals, for their own well-being and that of adjacent landowners, prevents them from moving on when their forage is depleted, thus allowing the range an occasional rest. Buffalo herds of a century ago were ruthlessly cropped back by nature's management tools — disease and starvation — when their numbers exceeded the capacity of the available range.

In a natural environment, all life is interrelated in a complex and intricate system, a system refined over millions of years by harsh trial and error. When one link is affected, that action must be balanced. The objective of the National Bison Range is to recreate a natural system in miniature, complete with its most intricate workings, tampered with only to ensure its perpetuation.

Man has learned much about how to manage a natural system, and today the National Bison Range grasslands are lush and healthy.
Michael Sample

Buffalo numbers are carefully balanced with available habitat on refuges like the National Bison Range. Jon Cates

This yellow-bellied marmot is in his species' classic pose—viewing the world from an observation deck high on a rock field. "Rockchucks" eat heavily all summer and hibernate from August until spring. Alan Nelson

Black bears are not full-time residents of the National Bison Range, but they sometimes climb the fences and pay a visit. They go into a period of dormancy but are not true hibernators. Jon Cates

Mule deer, so named because of their enormous ears, forage early in the day, retiring to shady, safe areas to regurgitate their food and chew it as cud. "Mulies" weigh 175 to 400 pounds and are best known for their distinctive hop, all four feet touching the ground at once.
Jon Cates

The National Bison Range is good habitat for the saw-whet owl, named for its call which sounds remarkably like the filing of a saw. The owl spends daylight hours hidden away in the dense conifer forest, waiting for darkness and the chance to hunt and feed on small nocturnal mammals.
Jon Cates

Pronghorns are not true antelope, despite their name, Antilocapra americana. *They are bundles of nervous energy that can blur across the plains at fifty-five miles an hour or lope along at thirty-five. Intensely curious, pronghorns often fell prey to Indians who lured them into range by lying in the grass, waving a rag on a stick.*
Michael Sample

Bobcats are rarely seen by visitors on the National Bison Range because of their artful ability to keep a low profile. Even kittens have the distinctive black ear tufts, which make fine antennas.
Jon Cates

Much as Michel Pablo's cowboys rode off to corral buffalo for shipment to Canada in 1907, modern wranglers begin the annual National Bison Range roundup. M.J. Gordon

Capturing bison is no easy task. The enormous and sometimes surly beasts don't submit willingly. M.J. Gordon

The dust flies just as high today as it did when Pablo's men were trying to persuade the buffalo to enter corrals.
Michael Sample

A strong herding instinct, originally developed to protect buffalo from predators, makes the October roundup a bit easier. M.J. Gordon

Ironically, this same instinct nearly drove the beasts to extinction. They were easy targets for hide hunters because they refused to disperse. Michael Sample

The yearly roundup is a public spectacle of large proportion. Reinforced corrals would pose no great challenge if the buffalo decided to make a break for freedom. Michael Sample

Blood samples are part of the routine. Jon Cates

A few buffalo submit to capture and are driven into a corral for sorting and physical exams. Michael Sample

Range manager Jon Malcolm moves a buffalo through the chutes.
M.J. Gordon

Other buffalo were selected for sealed-bid sale, but this bull will remain on the refuge for at least another year. M.J. Gordon

When bighorn sheep exceed the capacity of their forage on the National Bison Range, a few are transferred. Jon Cates

A helicopter flight is a traumatic but short-lived experience for a pair of bighorns selected for live delivery to a new location. Jon Cates

A CLOSER LOOK

The earliest written description of the American buffalo, by Spanish historian De Solis in the early 1500s, suggests the animal was pieced together from various parts of other creatures: ". . . it has crooked Shoulders, with a Bunch on its Back like a camel; its Flanks dry, its tail large, and its neck cover'd with Hair like a Lyon. It is cloven-footed, its head armed with that of a Bull, which it resembles in Fierceness, with no less Strength and Agility."

Technically, the buffalo should be called American bison. The great beasts actually are unrelated to the true buffalo of Asia and Africa. However, the term "buffalo" has been used for 150 years to describe the bison and even wildlife experts accept it.

Buffalo are the largest land animals to roam the North American continent in historic time. Mature bulls of nine or ten years measure nine to eleven feet from head to tail and may stand as high as six feet at the shoulder. Average weight is 1,400 to 2,000 pounds, but a few have been reported to weigh as much as 2,500 pounds. Cows are of a lighter frame and average between 700 and 1,100 pounds.

Cows on the National Bison Range live an average of twenty years, while the average lifespan for a bull is fifteen years because some are killed during the fierce battles of the rutting season. A few buffalo have been recorded to live forty years.

Though hulking beasts, buffalo are quite agile. Early-day explorers who encountered them in mountainous terrain noted incredible feats of climbing. Month-old calves can course the plains at speeds of thirty-five miles an hour for short distances. No refuges are large enough to measure a mature buffalo's endurance, but in the wild the animals were reported to maintain speeds in excess of thirty miles an hour for distances that a man on horseback could not match.

Few sights evoked a sense of power more than a running herd of buffalo, the ground trembling under the weight of their pounding hooves. In battle, they have been known to toss elk and even other buffalo several feet in the air, or over their backs.

For thousands of years, until man came along, the adult buffalo's greatest threat came not from predators, but from an environment in which temperatures varied as much as 140 degrees in a year, where wildfires raged unchecked over thousands of square miles, and prairie winds piled snow into drifts that buried small trees.

Buffalo appear top heavy because of their massive heads. So thick are their skulls that they retarded the bullets of hunters. Michael Sample

Herds have been reported to plod through drifts in which they were lost from sight. These same creatures are admirably suited to the plains' scorching summers. Though they drink water in prodigious quantities when they encounter a stream, they are capable of going for days with only the moisture derived from succulent grasses.

Creatures that pass the test of survival are not products of chance, but of natural selection. The buffalo was tailored precisely for life on North America's grasslands.

The buffalo's seemingly oversized head contributes to its popular image as clumsy and dull-witted. The thick skull and dense hair insulate the buffalo from severe prairie winters and provide armor during head-to-head combat between two bulls. This fortress of bone and fur defied the bullets of market hunters and guards the most important sense organs—eyes, nose and ears.

The buffalo's vision is not keen because gregarious herd life does not demand it, and lateral placement of the eyes precludes binocular vision. Nineteenth century Great Plains artist George Catlin wrote that "buffalo are very blind animals, and owing, probably, in great measure, to the profuse locks that hang over their eyes. . ." Such exaggerated beliefs about their poor vision probably were based on their habits of piling blindly over buffalo jumps or "pishkuns" when driven by Indian hunters, and of milling aimlessly while hide hunters shot dozens of them.

Hide hunters were quick to learn that the buffalo's sense of smell is acute. Even before a potential threat was seen or heard, entire herds would stampede off if the wind carried a scent of danger. Plainsmen of a century ago swore a herd could smell water for five miles or more, and locate a rich bunch of grass buried under a foot of snow.

The buffalo's ears are mere tabs nearly lost in a tangle of long hair, but its hearing is good because communication is an important part of herd social life: Bulls send and respond to bellowed challenges during the rut, and cows and calves use sound to locate and identify one another.

The pronounced hump, even more than the massive head, is a buffalo hallmark. Unlike the fatty humps of camels, the buffalo's is actually a feature of the spine, supported by elongated extensions of the vertebrae. Bulls' humps increase in size with age.

The function of the hump remains a subject of speculation. Behaviorists note that bulls typically arch their backs during confrontation to appear larger and more formidable to opponents.

(continued on page 39)

In 1843, John J. Audubon visited the Montana range and compared the bellowing confrontations of buffalo bulls to "the long continued roll of a hundred drums." Danny On

Buffalo calves are exceptionally hardy. Many were born during Michel Pablo's epoch roundup in 1906-1912, and survived the rigorous trip to Canada.
Jon Cates

With mere buttons for horns, calves weigh thirty to forty pounds at birth and within hours can keep pace with adults for short distances. Jon Cates

Buffalo have thick robes that made it unnecessary for them to migrate south in winter, even when freedom was theirs and they had the option.
Jon Cates

Some wildlife biologists suggest that the additional bone and muscle reinforces the neck and head during battle, and might provide extra strength for sweeping aside deep snow to reach grass.

The hump meat was highly prized by both Indian and white hunters. During the time of extravagant waste by buffalo hunters and pioneers, the humps were choice cuts, second only to the tongues. To many Indian tribes and French trappers, the strip of fatty tissue along the crest of the hump, which weighs five to eleven pounds, was the greatest delicacy.

Buffalo horns are not shed annually like antlers, but grow larger each year. The largest bull horns ever recorded spanned nearly three feet and measured sixteen inches around the base. Horns on older cows are more slender and curved.

These sharp-tipped horns are lethal weapons and no doubt brought sudden death to prairie wolves that pressed too close, grizzly bears, and many a favored hunting pony. During the savage battles of the rut, some bulls invariably are scarred by a challenger's horns, but fights to the death are the exception.

Considering the buffalo's thick winter coat, it seems remarkable that early observers believed buffalo herds of the Great Plains migrated south to escape winter. During autumn, the skin and fatty layers thicken and the sleek, nearly hairless buffalo of summer is coated with a dense mat of woolly hair. In autumn and winter, Indians hunted buffalo for robes, which they tanned with the hair on for use as groundcovers and wraps.

During the spring molt, matted clumps of winter fur are sloughed off by wallowing or by vigorous rubbing against rocks, trees, and, much to the dismay of early railroad men, telegraph poles. Frequent replacements were necessary and the addition of sharp spikes to discourage the itchy buffalo only made the poles more attractive.

Remnants of great herds that once filled the prairie from horizon to horizon, today's National Bison Range herds are small, carefully balanced with available forage.
Jon Cates

Buffalo use their heads like huge brooms to sweep away snow, burrowing as deep as necessary to find a morsel of prairie grass.
Jon Cates

Shortly after the turn of the century, when the American Bison Society was grasping for arguments in favor of bringing the species back from near extinction, it proposed that "buffalo wool" be spun into yarn. Gloves and a blanket made of the wool were warm and durable, but shearing the uncooperative and dangerous beasts proved an insurmountable obstacle to commercial production.

The buffalo's tail is an expressive feature of the animal's anatomy. During playful chases or while nursing, a calf's tail wags with pleasure. Among adults, the tail is down during moments of indecision, uncertainty and contentment, and is held high and switched from side to side during aggressive encounters, stampedes, and charges.

A closer look at the American buffalo reveals more than first meets the eye. Nature is a frugal draftsman, lavishing few frills on its creatures, yet seldom failing to provide the necessities. These shaggy beasts were wonderfully fashioned, their design refined by unrelenting winds, deep snows, searing heat and the vast expanses of the Great Plains.

Windswept ridges make foraging easier for grazing animals on the National Bison Range.
Jon Cates

Black-billed magpies scavenge for food on the open prairie but require trees for nesting. Michael Sample

Despite their docile appearance, bison are dangerous if approached too closely. Michael Sample

A male blue grouse displays by fanning his tail and inflating two neck pouches, producing a low, muffled hooting. The game birds prefer open forests and forest edges. Jon Cates

THE GRASSLAND YEAR

Arrowleaf balsamroot is an edible plant widely dispersed across Montana, including the National Bison Range. Jon Cates

SUMMER: The year of the buffalo begins

Evening light bathes a small rise on the edge of a sweeping basin where two buffalo bulls perform a ritual almost as old as the prairie itself. Twenty feet separate the enormous beasts. The five-year-old bull stands broadside, back arched and belly heaving, mouth agape and tongue lolling as he issues a bellowing roar.

The challenge is unmistakable. The larger, ten-year-old bull leaves the cow's side, paws the sun-dried soil, and wallows violently, his legs flailing in the air. Again on his feet, he snorts explosively and advances stiffly, stomping the ground, shaggy pantaloons bouncing.

Tail cocked high, the young challenger responds, swinging his head up and down, closing the distance between them. The young bull lunges and they meet head-to-head with such force that the old bull's forefeet leave the ground.

With awesome, brute strength they push and shove, their hooves scrambling for firm ground, tearing trenches in the prairie sod. In less than a minute the confrontation has ended. The old bull turns his head in submission, grabs a mouthful of grass to confirm his defeat, and walks away. The young bull has won admission to the ranks of breeding males and moves off to claim the cow.

It is July on the National Bison Range and tension mounts as the breeding season begins. Bachelor groups disband and bulls re-join the larger herds of cows and calves. Grooming activities assume new significance. Mock battles, half-hearted versions of the vicious battles to come, erupt with frequency.

During the rut, bulls establish and defend mobile territories, ten to twenty-five feet around cows, which first breed at two years of age and only tolerate the bulls' advances for two or three days at a time. If mating does not occur during this period, they will not become receptive again for about three weeks, accounting for the various sized calves seen in late summer.

With behavioral signals and by emitting a sexual scent, a cow signals her readiness to mate and the bull tends her, staying by her side and attempting to direct her away from other bulls. Only to meet a challenge or after the cow is no longer receptive will the tending bull leave to find another.

Buffalo bulls often spar with each other, but the battles become more serious as the mating season approaches, and occasionally end in death. Jon Cates

The clear flute-like song of the western meadowlark makes it a Montana favorite. Mottled coloration protects this ground nester from predators. Jon Cates

Summer forage is sheer luxury for bighorn sheep after a winter diet of dry grasses and forbs. Jon Cates

A fallen limb makes a fine scratching post for a bull elk that's come to drink, and admire himself. Jon Cates

If this female blue grouse remains still, chances are excellent a predator will never see it. Jon Cates

While the battles of rut are dramatic, dominance among breeding-age bulls is more often determined by elaborate but subtle behavior. Lethal confrontations are not to the long-term advantage of the species, so most challenges are settled by less violent pushing matches or ritualized displays and posturing. Even so, the cost of the mating season is high: A bull may lose as much as two hundred pounds during the month-long rut.

For most prairie life, July and August are months of quiescence, a pause between the growth and proliferation of spring and the labors of autumn in preparation for winter. The rush of spring wildflowers is buried in maturing stands of fescue, wheatgrass, Junegrass, and bluegrass. Coarse prairie forbs climb above the grasses and flower—golden aster, goldenrod, and gumweed, aromatic horsemint, and, as fall approaches, brilliant blue asters. The prairie is maturing. Ripening chokecherries hang in glossy clusters alongside hawthorn fruits in valley bottoms.

Summer is an abundant time when the young have swollen the numbers of every animal species. Deer are often seen in threes, as twin fawns trail behind does. Young pronghorns have outgrown their leggy awkwardness and lope smoothly in play. At higher elevations, bighorn lambs are gathered in nurseries under the watchful eyes of community mothers.

Small mammals reap and store the grassland's summer bounty. Voles harvest blades of grass and mice lay away seeds and fruit pits for the lean months of winter. Yellow-bellied marmots and Columbian ground squirrels have long since completed their parental chores and are swollen with layer upon layer of fat. For several weeks during the heat of summer, they estivate in cool underground burrows, in dormancy not unlike winter hiberation.

Some grassland birds have migrated south by the end of August, but most linger, feeding on a seemingly limitless supply of seeds and insects. Like the marmots, the birds build fat reserves, but not to sustain them through a deep winter slumber. Theirs will fuel a journey south to wintering grounds when the first hint of frost sweeps over the grassland.

AUTUMN: A Season of Preparation

Autumn comes early to the northern prairie. Grassland plants and burrowing animals retreat to their frost-free underground sanctuaries. For those animals that do not hibernate or flee to warmer climes, it is a time of preparation. Buffalo calves have lost their rusty summer coats and their backs are beginning to arch as their humps develop. Coats have thickened into dense blankets

Varied habitat is agreeable to the great horned owl, which helps keep nature's balance by feasting on rodents, hares, skunks and large birds including ducks, herons and even other owls. Jon Cates

unequalled in the animal world. The elk's short reddish-brown coat is replaced by a wooly brown undercoat and coarse, grayish guard hairs. Deer, pronghorns, bighorn sheep, and mountain goats gather insulating layers of hair and fat for the days when harsh winds will sweep down from the Arctic. Weasels have shed their summer browns for ermine-white.

On marshy ponds to the north of the Bison Range, mallard and redhead ducks, coots, Canada geese, and blue-winged teal are gathering for fall migration, joined every day by more of the same species from the north.

When the golds and browns of curing grass dominate the prairie, the mating season arrives for many grazing animals. By September, pronghorn bucks have begun to gather and defend large harems, even though mating is nearly a month away. Yearling bucks, now a threat to the older males, are driven away from the herds to wander alone over the grassland.

By late September, the breeding season of the elk is at a fevered pitch. Bachelor groups disband as aggression swells, and antler tips are polished to an ivory white for battle. The bugling of a bull elk is the epitome of all that is wild. Calls vary from roars to whistles but usually begin on a low note and climb, eerie and flute-like, before falling to a series of gasping grunts, a bull's way of issuing challenges and laying claim to as large a harem of cows as he can defend.

Deer come less conspicuously into rut in late October and early November. Bucks trail receptive does and occasionally lock horns with rivals for breeding privileges, mule deer with less intensity than whitetails.

The breeding season begins in November for mountain goats, when billies consort with one or two nannies, but do not gather harems. Intruding males are driven off, with bluffing but little serious fighting, in contrast to the last and most spectacular of all mating season battles—that of the bighorn sheep.

Only large rams of nearly equal horn size square off in fierce head-butting battles. Raising up on their rear legs and falling forward in powerful charges, the rams meet with such impact that splinters fly from their horns and a sharp report like that of a rifle echoes through the basins. Again and again they slam together until one acknowledges the other's dominance and drifts away, dazed.

Late fall seems like an imprudent time for the energy-robbing mating season but it has a certain logic. Fawning, calving and lambing must coincide with the abundance of the spring growing

(continued on page 51)

Golden pea adds color to the hillsides of the National Bison Range, but it's not a favorite of either livestock or wildlife. Michael Sample

Perhaps aware that winter brings short days and arctic winds, a bull buffalo soaks up all the sun he can during a summer afternoon snooze. Jon Cates

A whitetail races across the prairie, its tail acting as a warning flag for others. Jon Cates

The blanketflower or brown-eyed Susan adds its burst of color to the prairie. Michael Sample

Whitetails exchange their reddish summer coats for a darker winter coat that blends well with the somber landscape. Jon Cates

season, and so the gestation period dictates the time of rut. Rigorous battles and scant feeding deplete the males' fat reserves and lower their chances of surviving winter, but the females continue to feed heavily through the mating season. Most enter winter in peak condition, as they must, for they are the carriers of the next generation.

WINTER: A Season of Testing

Winter is an instrument of natural selection, culling the weak, old and poorly adapted, trimming the number of animals to fit the habitat. Each animal has evolved to cope with this most arduous of seasons in its own way.

Most insect-eating birds have no choice but to migrate south. Small birds that winter on the National Bison Range include Clark's nutcrackers, black-capped chickadees, nuthatches, Townsend's solitaires, Bohemian waxwings, evening grosbeaks, dark-eyed juncos and gray partridge. They subsist on seeds, fruits, and later, as winter relents, the swelling buds of trees.

Winged predators, like the rough-legged hawk, golden eagle, and short-eared owl eke out a living on small mammals. Scavengers like the black-billed magpie rely on the failure of other animals to endure the low temperatures, deep snow and reduced supply of food. Mallards, mergansers and goldeneyes linger as long as open water remains, often all winter, feeding on aquatic life or the leavings of neighboring grainfields.

For some animals, like the Columbian ground squirrel and the marmot, winter means a retreat underground, where their heartbeats, temperatures, and respiration are reduced to minimal levels that will sustain life and conserve fat reserves. Voles move above-ground with the first snow cover, constructing a maze of tunnels in the vegetation at ground level and building globular nests of dry grass. In this dark world they feed on plants, protected from the cold and concealed from predators.

Large grazing animals cannot escape winter, and the short days are spent in an unending search for food. Buffalo have given up their midday siestas. The cold does not burden buffalo, provided they have ample food. Their thick coats and layers of fat are efficient insulation. They are drawn to wind-blown ridges and flats where grasses are exposed and they can avoid deep snow, though they

Rodents and deer seek the yellow fritillary's tender seed pods and starchy corm, as did the Plains Indians. Jon Cates

can plod through drifts four feet deep, climbing on top with their forelegs and crushing it down. To reach the grass below, they push their noses into the snow and swing their heads from side to side to clear a trench.

As snows deepen, elk and bighorn sheep move down from the high country and join the buffalo on wind-cleared grasslands. Small bands of antelope merge into herds of 100 or more on the open plains.

Winter begins its retreat in early March and life on the National Bison Range is at its lowest ebb. In the most severe winters, small mammals deplete their food reserves and the large grazing animals are thinner from months of cold. Ironically, when relief is now near,' winter sometimes takes its greatest toll. A late winter storm that covers the grassland for an additional week will only further weaken some, but may bring starvation for others.

SPRING: Renewal of Life

In late March or early April, life begins to stir in the winter-bleached grass. Voles and deer mice give birth to the first of many litters, a crop on which great-horned owls will raise their own broods. Ground squirrels and marmots, eager to feed for the first time in nearly seven months, are above-ground to sun at midday.

The earliest and most delicate of the wildflowers bloom, hidden from view, hugging the ground, rushing to mature before taller grasses and forbs tower overhead. Many early flowers are of the iris and lily families, pushing up from a store of food in their starchy bulbs. Snows are still receding when the yellow fritillary sends up a single yellow flower on a drooping stalk. Seeds of blue-eyed Mary swell with moisture and germinate, and shooting stars appear alongside starflowers.

As new grass sweeps over the range, buffalo begin to calve, most in late April or early May. As the time of birth approaches, cows become restless and leave the company of the herd. Born a tawny orange and without humps, the calves, which weigh thirty to forty pounds, are licked dry and with wavering steps attend to the first order of business, nursing. Within hours, these same tottering creatures are at play, running and bucking around the cows.

As the calving season progresses, cows and calves band together in "maternity clubs," and by the time the young are two or three weeks old, they are spending as much time with each other as with their mothers. The bond remains strong, however, through the rigors of the summer mating season, until they are weaned in early winter.

The long-plumed avens is sometimes called "prairie smoke" and its bloom coincides with buffalo calving season. Indians used its root to make a tea. Michael Sample

In May and early June, deer and pronghorn are dro fawns, elk their calves, and bighorn sheep their lambs. buffalo calves, the young of these species depend on co for survival. Born essentially scentless, they are taught to re hidden for the first few days. True creatures of the grasslands, leggy pronghorn fawns soon find safety in speed. Deer fawns, li the adults, are secretive and only during the fading hours of the day or early morning leave the security of woody cover.

June is a month of sheer exuberance for plant and animal life on the National Bison Range. The population of large grazing animals is at its height, migratory birds have returned to nest, small mammals and insects have exploded in numbers and the delicate wildflowers of early spring have been supplanted tenfold with a rush of color that blankets every hillside and valley.

For so many plants to extract a living from one area they must distribute themselves in time and space. Early plants grow and flower quickly before the competition for light and moisture becomes keen. Later plants must climb higher to reach the sunlight. Each is adapted to a precise combination of moisture, soil fertility and light. Sticky geraniums, long-plumed avens and lupine colonize low valleys and north slopes. In more arid locations are pussytoes, yellow paintbrush, owl clover, penstemon and death camas. Yarrow, larkspur, fleabane, ragwort, blanketflower and balsamroot prefer higher, drier, south slopes. But nature abhors rules and hard lines. Members of these communities overlap and blend, shifting back and forth in wet and dry years.

Montana's state flower, the bitterroot, has little competition in the dry rocky soil of higher elevations. Before the snow has left, it sends up fleshy leaves to resupply its starchy roots. By the time its pinkish flowers appear, the leaves have withered. Having spread its seeds and replenished the food stored in its roots, the bitterroot retreats back into the earth before summer strips the soil of its moisture.

By the end of June, the antlers of the buck deer and bull elk are nearly full grown. Short-eared owls and hawks are cropping the populations of small mammals, and buffalo bulls are once again pawing, wallowing and sparring. The rutting season, the beginning of another biological year on the National Bison Range, is fast approaching.

Streamside transition zones provide excellent cover for the showy wood duck. Jon Cates

These prickly seed pods of teasels, atop tall stalks, stand drying in the autumn sun. Michael Sample

Birds depend on the seeds of the Mullein plant, and if forage is scarce, elk will eat the stalks and leaves. The leaves yield a substance that can be used to soothe inflammation. M.J. Gordon

Buffalo are well-adapted to the harsh prairie winter, and the cold poses no problem for them if forage is available. Jon Cates

Late afternoon sun bathes the Mission Mountains in a cold winter light. Michael Sample

Mule deer bucks lock horns in a friendly sparring match not meant to seriously harm. Jon Cates

The pugnacious badger spends most of its life burrowing after rodents, snakes and hares. It may hole up and eat for several days after an especially succcessful hunt. Jon Cates

Newborn fawns rely on their speckled camouflage and their lack of scent for protection after birth. Jon Cates

Bighorn ewes take turns "babysitting" the crop of newborn lambs in community nurseries. Jon Cates

Open forests are perfect habitat for the arnica, a potent medicinal plant that can be used to increase body temperature or reduce infection. Michael Sample

Calves become independent soon after birth and wander away from their mothers for long hours of play. Michael Sample

A young beaver finds plenty to gnaw on in deciduous streamside woodlands (right). Jon Cates

The adaptable robin, forerunner of spring, is equally at home in country and town (far right). Alan Nelson

GEOLOGY & HABITATS

Clarkia is a member of the evening primrose family, named for its Lewis and Clark expedition discoverer. Michael Sample

The tranquility of the National Bison Range belies the cataclysmic forces that carved the landscape. Violent upheavals tossed about sheets of rock miles thick, while other forces methodically ground away at the land for millions of years.

For more than a billion years, the northern Rocky Mountain region was a vast, shallow sea that accumulated sediments to a depth of over 30,000 feet, deposits that would consolidate into limestones, mudstones, and sandstones over which bighorn sheep would one day scramble.

One hundred million years ago, this geologically placid setting began to tremble with impending change. Hot masses of semi-solid rock began to rise under the earth's crust, a skin only 25 miles thick. Not in a single convulsion, but in thousands of smaller, violent shudders, huge blocks of rock were shoved upward along a fault line. The Rocky Mountains were rising.

The land tilted and what remained of the vast inland sea became a dumping basin for debris and erosion outwash from the weathering mountains. The climate was dry for about 40 million years after the initial period of mountain-building, so few streams carried away the sand, gravel, and mud that had accumulated to thousands of feet deep in the valleys.

Three million years ago, a mere flicker of geologic time, the National Bison Range and Flathead Valley would not have been recognizable. The final sculpting was accomplished by yet another great geologic force. The Pleistocene, or Ice Age, was not a period of intense cold as is commonly believed, but rather a time when snowfall accumulated faster than summer's heat could melt it. Four times during this epoch, great ice sheets crept south over Canada and much of the northern United States. Each time they cut valleys deeper, making the mountains seem higher, chiseling deep cirques in the sides of the mountains, and establishing river systems that today continue to carve into the valley fill.

When the mammoth glaciers halted and began to shrink back, they left behind immense mounds of rock, rubble and blocks of ice. Once such terminal moraine lies north of the National Bison Range near Ninepipe National Wildlife Refuge. Another impounded Flathead Lake. As the ice melted, hundreds of glacial kettles or potholes were formed and today are ponds and wetlands vital to wildlife.

A Columbian ground squirrel feeds on the prairie's summer abundance, gaining layer upon layer of fat to sustain it through the winter. Jon Cates

During the final ice advance, a glacier dammed the Clark Fork River in northern Idaho, creating Glacial Lake Missoula, which flooded many of the valleys of western Montana. The National Bison Range was an island in one arm of this expansive lake, and beach lines are still visible on the north slope of Red Sleep Mountain, the highest point on the refuge. Only ten thousand years have passed since the last sheet of ice withdrew to the north.

Less spectacular forces continue to change the face of the National Bison Range. Streams cut deeper into the valleys, tumbling and grinding rocks to pebbles, pebbles to grains of sand, sand to silt. Freezing and thawing expands fractures in the rock, and wind carries away rich soils. The land seems forever and the trees and grasses, mountains and streams, buffalo and bighorn sheep have found their niches, but they are only part of a fleeting moment in geologic time.

It is apparent at first glance that the National Bison Range is both forest and grassland, and indeed, these are the most extensive habitats. But within them are others, and within each are micro-habitats that meet the precise requirements of one plant or animal species.

The grassland is one of the last and most pristine examples of Palouse prairie, a type of range that once occurred throughout the Flathead Valley, northern Idaho, Washington and Oregon. Most of it has been converted into wheat fields or transformed into sagebrush flats by overgrazing.

In its natural condition, Palouse prairie is an open yet luxuriant assemblage of Idaho and rough fescue on less-arid north slopes; bluebunch wheatgrass, Junegrass and Sandberg bluegrass on drier south slopes. Among the grasses, a diverse collection of forbs and showy wildflowers flourish — balsamroot, yarrow, paintbrush, larkspur, fleabane, sticky geranium, lupine and others.

Though wedged between mountain ranges that have influenced the nature and composition of the plant life, Palouse prairie exhibits characteristics common to all grasslands—rolling or flat terrain, low precipitation with hot, dry summers, and periodic drought. While these factors of climate and topography demand that grasses be the dominant vegetation, other elements of the environment have made it a unique grassland.

Because of its geologic history, the soils of this intermontane valley prairie are unusually fertile, like the tallgrass region of the eastern plains. Precipitation, though scant, comes primarily as snow, allowing deep percolation into the soil before summer heat can

The American bittern points its neck and head skyward and remains still when predators are near. Thus it blends with the wetland's vegetation. Jon Cates

evaporate it. Summer days are hot but nights are cool, and absent from this "mountain prairie" are the unrelenting winds that desiccate open plains.

The National Bison Range's northern latitude favors grasses which initiate growth and bear seeds in a cool climate with a short growing season. And so, in this land of low rainfall, the Palouse prairie evolved, an unexpected lush grassland in a mountainous region.

Though perhaps not as inspiring as a sequoia or showy as a wildflower, grasses are wonderfully adapted to their environment. Rather than resisting the ubiquitous prairie wind, grasses are supple, bending with it, even employing it to spread their pollen and seeds. The leaves are long and slender, exposing a minimum of surface to drying winds.

More than half of a typical grassland plant is underground, where deep taproots or a fibrous network of roots absorb every available drop of moisture. The growing tissue is near ground level so grazing animals and raging wildfires do not kill the plant. New leaves simply push up from the rich prairie soil.

With palatable grasses and forbs so abundant, grazing animals are the dominant life of the prairie. Just as plants transform solar energy and elements of the earth into a living, renewable resource, these herbivores convert plants into animal tissue.

Buffalo, elk, pronghorn, deer, and large numbers of rodents and insects all graze or browse on prairie plants. They co-exist harmoniously because each is a specialist. Some competition occurs, but there is a natural sorting by food and habitat preference. Buffalo feed principally on grass, pronghorn on forbs and sagebrush. Elk also feed primarily on grass, but compete directly with buffalo only when the snow drives them down from the mountains. White-tailed deer are browsers but prefer woodland edge and are usually found at lower elevations. Mule deer require less woods or shrubs than the whitetails but avoid the open, flat grasslands used by pronghorn.

The grasslands support a multitude of more secretive plant eaters, collectively the most voracious consumers of the prairie. The smallest are the insects, particularly the many varieties of grasshoppers, caterpillars and beetles.

In this prairie land, life seems to exist in a shallow zone between the soil and the hill's tallest grasses. Over thousands of years, grassland animals have adapted well. The dominance of burrowing and grazing animals is characteristic of every grassland on

The sharp ebony horns of Rocky Mountain goats can become lethal weapons, but the agile animals prefer to simply outclimb their predators. Jon Cates

Earth. On the Palouse prairie the burrowing mammals are represented by the yellow-bellied marmot, the Columbian ground squirrel, and the northern pocket gopher.

The Palouse prairie does not support a large variety of birds, but those species that have adapted are abundant. Of necessity, nearly half nest on the ground and the others are seldom far from it, raising their young in stiff weeds, clumps of grass or shrubs.

Ground nesters such as sparrows, meadowlarks, nighthawks, and killdeer are particularly vulnerable to predators, and so most wear a mottled camouflage to protect themselves. Visibility is not restricted on this open grassland, so bright advertising colors are not required.

Native grassland covers more than two-thirds of the National Bison Range, but three other major habitat types also are found within its boundaries. The most visible is the fir-pine forest found at higher elevations. Douglas fir dominates north slopes where moisture is more abundant, while ponderosa pine tolerates drier conditions found on south slopes and higher elevations. Its spreading root system so efficiently taps the soil of its moisture that these woodlands are typically quite open, with evenly spaced trees and little undergrowth of grasses or shrubs.

Mammals closely associated with the National Bison Range montane forest include elk, mule deer, bobcats, yellow pine chipmunks, red squirrels, bushy-tailed woodrats, long-tailed voles, porcupines, and snowshoe hares. Black bears occasionally enter by climbing the fences, but they seldom take up residency on the refuge.

Birdlife characteristic of the shadowy world of pine and fir includes blue grouse, Lewis' and pileated woodpeckers, Steller's and gray jays, Clark's nutcrackers, nuthatches, grosbeaks, pine siskins, and red crossbills.

Less than a square mile of rocky outcrop and talus slope is found on the National Bison Range, but it is essential habitat for mountain goats and bighorn sheep. Even here, there is a sorting by food and terrain requirements. The goats seldom stray far from the most rugged sites, plucking whatever bit of greenery they can from cracks in the rock. Sheep demand more palatable plants and so frequent lofty, open slopes or basins near rock walls or outcrops to which they can escape if threatened by predators. Both species are equipped with rough pads on their hooves, spreading toes, and an amazing sense of balance and confidence that allows them to bound over steep rock faces.

The greatest number and diversity of wildlife species on the National Bison Range are found on the forest edge and in the streamside thickets, transition zones that support a gathering of plants from both woodlands and grasslands, as well as others not found in either. Food, cover, nesting and denning sites, water, singing perches, and other needs all are found here.

The site for the National Bison Range was selected principally for its quality grasslands, but the other habitats have made it more than just a buffalo refuge. For its small area, the National Bison Range contains a great diversity of plant and animal life, the product of chance when powerful geologic forces sculpted the face of western North America.

Elk antlers are grown each summer, nourished by a thin layer of "velvet" that is sloughed off in time for autumn rutting season battles.
Jon Cates

Summer range offers lush nourishment and a respite from the harsh days of winter. Michael Sample

The range turns golden as summer recedes. Jon Cates

Black-tipped ears and dark eyes are the only clues to this snowshoe hare's location if it remains still. Michael Sample

At eight days, a buffalo calf has filled out and begun to resemble the adult. Jon Cates

Wetlands are important habitat for a variety of birds and animals on the National Bison Range. Jon Cates

An elk calf ten hours old is odorless and still, its spotted back helping it to remain inconspicuous in a sun-dappled clearing.
Jon Cates

An alert coyote has spotted a mouse and is ready to pounce.
Michael Sample

Bitter cold doesn't penetrate the hollow-haired coat of a white-tailed deer. Jon Cates

Mallards spend summers on the National Bison Range but migrate south when the potholes begin to freeze. Jon Cates

A clump of lupine provides excellent cover for this little cottontail.
Michael Sample

A solitary bull elk is one of about 100 or 150 of his kind on the National Bison Range. Jon Cates

National Bison Range employees occasionally ride the refuge to check on the animals.
Jon Cates

Buffalo relax in the warmth of summer. Jon Cates

Roundup corrals are empty until next year. Jon Cates

Golden eagles are the largest raptors on the National Bison Range. An adult may stand forty inches tall and sport a wing span of seven and a half feet.
Alan Nelson

Once nearly extinct, the American bison returned to claim its place as grasslands king. Jon Cates